The PMO Leader's Guide to Agile Transformation

Driving Innovation & Value in

Project Management

ISBN: 979-8-9908793-1-7
Printed in USA
Cover Design: DesignbyDutchess.com
First Edition

Published by SHO-TEK LLC

www.sho-tek.com

Contents

Chapter 1: Embracing Change and Driving Innovation

The Changing Landscape of Project Management

In today's fast-paced, highly competitive business environment, organizations face unprecedented challenges and exciting opportunities. To navigate the complexities and uncertainties of this landscape, traditional approaches to project management need to evolve. The predictable, plan-driven methods that served us well in the past are no longer sufficient. Organizations must embrace a more flexible, adaptive, and innovative approach to delivering projects and driving organizational success to thrive amidst disruption.

As project management professionals, we find ourselves at an inflection point. We must find a way to avoid clinging to outdated, bureaucratic practices that bog down our organizations and impede our ability to deliver value. The risk of falling behind our more agile competitors, missing opportunities for growth, and failing our stakeholders is too high. As forward-thinking leaders, it is up to us to champion a new paradigm for the project management office (PMO) that enables our organizations to harness change and unleash innovation.

The Evolving Role of the PMO

Traditionally, the PMO is the guardian of project management standards, methodologies, and governance. These are certainly vital responsibilities that will continue to play a key role. However, to maintain relevance and add value in an agile world, the PMO must broaden its purview and embrace new ways of working.

Rather than acting primarily as an enforcer of rigid processes, the PMO becomes an enabler and catalyst for organizational agility by adopting agile practices at a project team level and helping to shape an agile culture and mindset across the enterprise. It means becoming a strategic partner to the business, working collaboratively with stakeholders at all levels to align project outcomes with organizational goals. Fundamentally, it means shifting from a control posture to empowerment, servant leadership, and continuous improvement.

This evolution will take time, and every organization's path forward is different. However, there are some common shifts that we as PMO leaders must begin to make if we wish to navigate the transition successfully:

- From rigid adherence to process to supporting flexible, adaptive practices.
- From a focus on deliverables and milestones to enabling incremental value delivery.
- From making long-term, detailed plans to facilitating iterative planning cycles.
- From a command and control mindset to fostering self-organizing teams.
- From managing tasks and resources to empowering people and removing impediments.
- From status reporting and oversight to facilitating transparency and continuous improvement.

The journey to agility is challenging, but for PMOs that embrace the challenge, the rewards can be significant. By adopting an agile mindset and culture, we position ourselves to help our organizations thrive amidst disruption and uncertainty. We enable faster time-to-market, increased responsiveness to change, improved stakeholder satisfaction, and better business outcomes. Most importantly, we create an environment where creativity and innovation flourish, and project teams find meaning and fulfillment in their work.

Becoming a genuinely agile PMO is an ongoing process that requires leadership commitment, hard work, and a willingness to learn and adapt. This book will explore key concepts, practices, and case studies that will guide you along your agile transformation journey. But first, let us take a closer look at what it means to be an agile PMO and the benefits this approach can yield for your organization.

What Will You Do?

As a PMO leader, project manager, or executive, you have a vital role in leading your organization toward greater agility. Consider taking these initial steps to begin your learning journey:

🎯 PMO Leaders: Invest time in understanding the Agile Manifesto and its twelve fundamental principles. Explore how they might apply in your organizational context.

🚀 Program Managers: Identify a pilot program or project to begin applying agile practices. Start small and focus on incremental learning and improvement.

📈 Project Managers: Educate yourself on the key agile project management frameworks Scrum, Kanban, and Lean. Practice applying them in your daily work.

💼 Executives: Engage your PMO leadership in a dialogue about your current project and portfolio management practices. Discuss the potential benefits of an agile approach.

👑 Chief Executives: Reflect on how well your organization's culture and leadership styles support agility. Identify opportunities to lead by example and drive change from the top.

Every organization's path to an agile PMO is unique, but it begins with a willingness to embrace a new way of working. By taking small, incremental steps and continuously learning and adapting, we can steer our organizations toward a more responsive, innovative, and value-driven future. The journey starts now.

Chapter 2: Understanding Agile Principles and Practices

To effectively lead an agile transformation within your PMO and organization, it is essential first to understand the principles and practices that underpin the agile approach. While agile methodologies are rooted in software development, the values, and concepts they embody apply across industries and business functions. This chapter will explore the Agile Manifesto, the key frameworks you should know, and the mindset shift required to embrace agility successfully.

The Agile Manifesto: A Guiding Light

The Agile Manifesto[1], created in 2001 by software development leaders, remains essential for understanding agile values and principles. While the specific implementations of agile have evolved over the past two decades, the core ideas espoused in the manifesto continue to inform and inspire practitioners worldwide.

The manifesto has four critical value statements:

1. Individuals and interactions over processes and tools

2. Working software over comprehensive documentation

3. Customer collaboration over contract negotiation

4. Responding to change over following a plan

These values don't imply that things on the right side of the statements (e.g., processes, documentation) are unimportant. Instead, they assert that we should prioritize and value the items on the left more. For the PMO, embracing these values requires a significant shift from a heavy focus on compliance, documentation, and rigid planning toward a more people-centric, collaborative, and adaptive approach.

Underlying the four values are 12 guiding principles that provide more specificity on what it means to be agile. These touch on concepts such as welcoming changing requirements, delivering working software frequently, face-to-face communication, sustainable development pace, technical excellence, simplicity, and self-organizing teams. While not

---

[1] https://agilemanifesto.org/

prescriptive, these principles offer guideposts for the PMO looking to enable greater agility.

A deeper dive into a couple of the principles highlights some of the crucial mindset shifts the PMO needs to make:

Customer satisfaction is paramount. For the PMO, this means moving away from a project-centric view focused on delivering a defined scope on schedule and budget toward a product-oriented view aimed at continuously delivering value to customers in small increments. It requires close, ongoing collaboration with business stakeholders and a willingness to use feedback and learning to adjust plans.

Use the momentum created by motivated individuals. Create a barrier-free environment to inspire teams to get work done. This principle challenges command-and-control management styles and instead emphasizes servant leadership. For the PMO, it means focusing less on assigning tasks and managing timelines and more on empowering skilled, self-organizing teams. It means providing the necessary resources, removing blockers, and getting out of the way.

While PMOs can dogmatically implement only some of the 12 agile principles, internalizing their spirit and intent is crucial for leading an effective agile transformation. The manifesto and its principles should serve as a constant reminder and guide for the PMO looking to enable true organizational agility.

Agile Frameworks: A Toolbox for Practitioners

With the values and principles of the Agile Manifesto as a foundation, numerous methodologies and frameworks have emerged to provide structure and practice for implementing Agile. While each has its unique flavor and areas of emphasis, they share common roots in the manifesto and can be valuable toolboxes for the PMO.

Some of the most widely used and recognized agile frameworks include:

Scrum is a popular methodology used extensively in software development and increasingly in other domains. Self-organizing teams, time-boxed iterative cycles called sprints, daily stand-up meetings, sprint reviews, and retrospectives characterize it. Scrum emphasizes transparency, inspection, and adaptation to improve the product and the team's practices continuously.

Kanban is a workflow management method that emphasizes visualizing work, limiting work in process (WIP), and optimizing flow. Teams using Kanban rely on visual boards to represent work items flowing through different process states (e.g., To Do, Doing, Done). It is less prescriptive than Scrum and can be a good choice for teams doing continuous production support or maintenance work.

Lean, which originated in manufacturing, has significantly influenced the agile movement. It is centered around maximizing value and minimizing waste. Key lean concepts include optimizing the whole, eliminating waste, building quality, deferring commitment, delivering fast, respecting people, and continuous improvement.

Extreme Programming (XP) is an agile methodology focused primarily on software engineering practices. It emphasizes close collaboration between developers and customers, short development cycles, continuous testing and integration, and evolutionary design. Typical XP practices include pair programming, test-driven development, continuous integration, and regular refactoring.

As a PMO leader, your role isn't necessarily to be an expert practitioner in all these methods. However, developing a working knowledge of the frameworks and their core practices is essential. You'll need this foundation to effectively support and guide your project teams in applying agile practices and to lead the broader organizational transformation.

It's also important to recognize that most organizations don't apply these frameworks in a purist fashion. Instead, they adapt and blend practices from multiple methods to suit their unique context and needs. As a PMO, your goal should be to enable a flexible, fit-for-purpose approach to agile rather than rigidly enforcing a particular methodology.

The Agile Mindset: A Shift from Doing Agile to Being Agile

While understanding the principles and practices of agile is crucial, it's necessary to recognize that true agility is more than a set of processes or tools. At its core, agile represents a fundamentally unique way of thinking about and approaching work. Embracing an agile mindset is just as important, if not more so, than adopting any specific framework.

Some of the fundamental mindset shifts that characterize an agile approach include:

From certainty to discovery: Traditional project management often assumes that project teams should fully define requirements upfront. Agile acknowledges that we must allow for emergence and discovery in a complex, rapidly changing world. We plan and execute work in smaller chunks, learning and adapting.

From control to empowerment: Command-and-control management relies on centralized decision-making and rigid hierarchies. An agile mindset values decentralized, self-organizing teams empowered to make decisions and given the autonomy to find creative solutions.

From predictability to adaptability: Detailed, long-term plans offer a comforting sense of predictability but often fail to survive contact with reality. Agile favors shorter planning horizons, a focus on iterative delivery, and the ability to pivot when conditions change, or new information emerges.

From risk avoidance to experimentation: In traditional project management, failure is seen as something to avoid at all costs. An agile mindset embraces experimentation and views failures as opportunities for learning and improvement. The emphasis is on "failing fast," learning from experience, and continuously improving.

From silos to collaboration: Conventional organizations often operate in functional silos, with teams working independently on their piece of the puzzle. Agile breaks down these walls through cross-functional teams, close collaboration with stakeholders, and a focus on optimizing the whole rather than micro-managing the parts.

From documentation to conversation: While not discounting the importance of key artifacts, agile values face-to-face communication over detailed documentation. The emphasis is on creating a shared understanding through dialogue and collaboration rather than producing reams of paperwork that quickly become outdated.

For the PMO, cultivating an agile mindset means leading by example and embodying these values in your work. It means fostering an organizational culture that supports experimentation, empowerment, and continuous learning. It means coaching leaders at all levels to let go of command-and-control habits and to trust and enable their teams.

This mindset shift is often one of the most challenging aspects of an agile transformation but also one of the most crucial. Without a fundamental change in thinking, agile risks becoming just another set of processes to follow rather than a true transformation in how work gets done.

What Will You Do?

Developing a deep understanding of agile principles, practices, and mindset is a continuing journey. Consider these next steps in your learning path:

🎯 PMO Leaders: Facilitate an interactive workshop with your team to explore the Agile Manifesto values and principles. Discuss what each would mean in practice in your organization.

🎯 Program Managers: Evaluate your current program management practices against agile principles. Identify areas where you could experiment with more agile approaches.

📈 Project Managers: Pick one agile framework to dive deep into, whether Scrum, Kanban, or another. Read a book, take a course, or find a mentor to expand your practical knowledge.

💼 Executives: Schedule a series of conversations with your PMO leader to understand their perspective on the potential benefits and challenges of adopting agile principles in your organization.

👑 Chief Executives: Reflect on your leadership style and mindset. Consider where you may need to adapt your approach to model and enable agility in your organization.

Embracing agile is not about blindly following a prescribed set of practices but about internalizing the values and principles and applying them in a way that works for your unique context. By continually deepening your understanding and challenging your assumptions, you lay the foundation for a successful agile transformation.

Chapter 3: Transforming the PMO to Embrace Agility

With this foundational knowledge of agile principles and practices as our guide, we now focus on the practical steps in transforming a traditional PMO into one that enables and supports organizational agility. This journey involves assessing the current state, crafting a vision for the future, navigating change challenges, and engaging stakeholders.

Assessing Your PMO's Agile Readiness

Before embarking on an agile transformation, it's essential to honestly assess your PMO's current practices, capabilities, culture, and mindset. This baseline assessment will help you understand your starting point, identify areas for improvement, and inform your transformation roadmap.

Some key areas to evaluate include:

Project Management Methodologies: What frameworks and processes does your PMO currently use to manage projects? How closely do they align with agile values and practices? Are they consistently applied across the organization?

Team Skills and Capabilities: How experienced and knowledgeable are your project teams in agile methods? Do they have the necessary technical and soft skills to work in an agile way? What training and support do they need to be successful?

Culture and Mindset: How well does your current organizational culture support agile values such as collaboration, transparency, experimentation, and continuous improvement? Are leaders at all levels bought into the agile mindset?

Governance and Decision Making: How are projects governed and decisions made? Is there a heavy emphasis on upfront planning and rigid controls, or is there room for adaptation and emergent learning?

Performance Metrics: What metrics does your PMO currently use to measure project success? Are they primarily focused on scope, schedule, and budget, or do they consider factors such as value delivery, customer satisfaction, and team morale?

Conducting this assessment can involve a variety of methods, such as:

- Interviews and focus groups with project teams, stakeholders, and leaders.
- Surveys to gather broad input on current practices and pain points.
- Process mapping to visualize and analyze current workflows.
- Benchmarking against industry best practices and agile maturity models.
- Agile readiness assessments to gauge organizational preparedness.

The goal is not to achieve perfection before your transformation but to clearly understand your current state, the gaps needing attention, and the strengths you will augment.

Crafting Your Agile PMO Vision and Roadmap

With a baseline assessment, the next step is to articulate a clear and compelling vision for your future agile PMO. This vision should paint a picture of how your PMO will operate differently, the value it will deliver to the organization, and how it will support teams in achieving better outcomes.

Elements to consider in your vision:

Agile Practices: What specific agile methodologies and practices will your PMO adopt and promote? How will you tailor them to fit your organizational context?

Roles and Responsibilities: How will the roles and responsibilities of the PMO and project teams change in an agile model? What new roles (e.g., Agile Coach, Product Owner) will you introduce?

Governance and Metrics: How will you adapt your governance model to enable more autonomy and faster decision-making? What new metrics will you use to track progress and success?

Organizational Structure: Will you need to restructure teams or reporting lines to enable greater agility? How will you foster cross-functional collaboration?

Culture and Mindset: What cultural shifts will be necessary to support an agile way of working? How will you promote trust, transparency, and continuous learning?

Your vision should be aspirational but achievable, forward-looking but grounded in your organization's realities. Most importantly, it should inspire and motivate stakeholders to join you on the journey.

Once you have a clear vision, the next step is to translate it into an actionable transformation roadmap. This roadmap should break down the journey into manageable phases, each with specific goals, milestones, and deliverables.

A typical transformation roadmap might include phases such as:

- Pilot and Learn: Start small by piloting agile practices with a few selected teams. Use these early experiences to learn, iterate, and build momentum.

- Scale and Standardize: Begin expanding agile practices to more teams, while developing common standards, tools, and processes to enable consistency.

- Optimize and Improve: Focus on continuous improvement by fine-tuning your practices, addressing challenges, and measuring results.

- Embed and Sustain: Institutionalize agile as the new standard way of working while continuing to evolve and adapt over time.

Your roadmap's specific phases and timeline depend on your organization's size and complexity, the urgency of the need for change, and your available resources.

As you develop your roadmap, remaining flexible and adaptable is essential. Recognize that things may go differently than planned and prepare to pivot based on learning and feedback.

Overcoming Resistance and Challenges

Any significant organizational change comes with its share of challenges and resistance. An agile transformation is no exception. People may be skeptical about the benefits of agile, fearful of how it will impact their roles, or simply resistant to changing the status quo.

As a PMO leader driving an agile transformation, it's crucial to anticipate and proactively address these challenges. Some common obstacles you may encounter include:

Lack of Leadership Buy-in: An agile transformation is unlikely to succeed without strong support and sponsorship from senior leaders. You'll need to engage executives early and often, building a compelling case for change and demonstrating how agile aligns with strategic objectives.

Change Fatigue: In organizations that have undergone multiple change initiatives, people may be weary of yet another "flavor of the month." It's important to acknowledge this context and position agile not as a fad but as a fundamental shift in how work gets done.

Skill and Knowledge Gaps: Agile practices could present a steep learning curve for some. Investing in comprehensive training, coaching, and on-the-job learning opportunities can help ease this transition.

Organizational Silos: Agile breaks down barriers between functions and encourages cross-team collaboration, which can be threatening to those used to working in silos. Promoting a "one team" mindset and creating opportunities for cross-functional working can help overcome this challenge.

Compliance and Regulatory Concerns: In heavily regulated industries, there may be concerns about how agile practices comply with standards and controls. Partnering with legal, risk, and compliance teams early in the process will help navigate these issues.

Addressing resistance and challenges requires a combination of strategic and tactical interventions. Communicating early, often, and transparently about the rationale, benefits, and impacts of the change, along with actively involving stakeholders in shaping the transformation journey rather than imposing top-down change will help keep teams engaged.

With the PMO and executives leading by example, visibly modeling agile values and behaviors, the transformation weaves into the organizational fabric.

Remind yourself and your teams to embrace the journey and focus on progress. Embedding agile within your organization will take time, persistence, and resilience.

Engaging and Empowering Stakeholders

An agile transformation is not the job of the PMO alone. It requires active engagement and participation from various stakeholders across the organization.

As ultimate decision-makers and culture setters, engaging and securing support from senior leaders is critical. They must understand and actively champion the vision, benefits, and their role in supporting the transformation. Project and program managers serve on the front lines of project delivery, and the shift to agile most directly impacts these individuals. Engaging them early, addressing their concerns, and equipping them to lead with an agile mindset is crucial. Agile often involves changes to traditional reporting lines and team structures. It is essential to engage functional managers in defining their new roles and ways of working.

Team Members: The individuals doing the day-to-day work of projects need to feel bought into the agile approach. Soliciting their input, providing training, and empowering them to shape new ways of working can help build grassroots support. Ultimately, agile is about more effectively delivering value to customers. Engaging them in the journey, seeking feedback, and demonstrating tangible results can help build external advocacy.

Effective stakeholder engagement is not a one-time event but an ongoing process throughout the transformation journey. Some tactics to consider include:

- Conducting stakeholder interviews and focus groups to understand needs, concerns, and ideas.
- Establishing an agile transformation steering committee with cross-functional representation.

- Hosting town halls, webinars, and roadshows to communicate progress and seek feedback.

- Providing targeted training and coaching for different stakeholder groups.

- Celebrating successes and recognizing individuals and teams who exemplify agile values.

The goal is to inform stakeholders and truly empower them as active participants and co-creators in the transformation process.

What Will You Do?

As you embark on your PMO's agile transformation journey, consider the following steps:

> 🎯 PMO Leaders: Conduct an honest assessment of your current practices and readiness for agile. Use this to inform your vision and roadmap.
>
> 🏹 Program Managers: Identify a pilot project or initiative to begin experimenting with agile practices. Capture learnings to inform broader rollout.
>
> 📈 Project Managers: Seek out opportunities to develop your agile skills and knowledge, whether through training, reading, or mentoring.
>
> 💼 Executives: Engage your PMO leader in a discussion about the potential benefits and challenges of an agile transformation. Consider how you can visibly support the effort.
>
> 👑 Chief Executives: Reflect on your organization's appetite and readiness for an agile transformation. Work with your leadership team to build alignment around a shared vision.

Transforming to an agile PMO is not a small undertaking, but the benefits of faster delivery, greater adaptability, and more engaged teams can be significant. By starting with a clear assessment, crafting a compelling vision, proactively addressing challenges, and engaging stakeholders along the way, you can set your PMO up for a successful agile journey. The path ahead may be challenging, but the destination - a genuinely agile organization - is well worth the effort.

Chapter 4: Agile Portfolio Management

As the PMO transforms to embrace agility, one of the most significant shifts is how we approach portfolio management. Traditional portfolio management often focuses on annual planning cycles, fixed budgets, and a command-and-control approach to project oversight. Agile portfolio management, in contrast, emphasizes continuous planning, flexible funding, and decentralized decision-making to enable faster adaptation to change and more dynamic allocation of resources to the highest-value work.

In this chapter, we'll explore the principles and practices of agile portfolio management and how the PMO can lead the transition to this new way of working.

Aligning the Portfolio with Agile Principles

At its core, agile portfolio management is about the strategic application of agile principles, including:

Rather than locking into a fixed annual portfolio, agile organizations continuously adjust their mix of investments based on the latest information about market conditions, customer needs, and business priorities. Agile portfolios should be flexible and adaptive, with mechanisms for quickly reallocating resources based on emerging opportunities or risks. Rather than micromanaging projects, agile portfolio management trusts and enables teams to make decentralized decisions about delivering value within strategic guidelines best. Agile portfolios involve ongoing engagement with business stakeholders to ensure investments align tightly with customer needs and value.

The shift to agile portfolio management requires more than just a process change - it demands a fundamental mindset shift about how we plan, fund, govern, and measure the success of our project investments.

Implementing Lean and Agile Portfolio Practices

So, what does agile portfolio management look like in practice? While specific approaches may vary, some standard practices include implementing Lean Portfolio Management (LPM). LPM is a set of principles and practices for applying lean and systems thinking to optimize the flow of value across the portfolio.

Elements of LPM include:

- Defining strategic themes and value streams to organize and prioritize work.

- Using lean budgeting and guardrails to fund value streams rather than projects dynamically.

- Establishing portfolio Kanban systems to visualize and manage the flow of work.

- Applying lean governance practices to enable decentralized decision-making.

Adopting agile funding models means shifting to more flexible and adaptive funding approaches rather than funding projects with fixed annual budgets.

Standard agile funding models include:

- Rolling-wave budgeting, where funding is allocated in shorter-term increments based on the latest priorities and performance data.

- Value stream funding, where long-lived, cross-functional teams receive funding to deliver ongoing value in a specific business area.

- Capacity-based funding, where teams work within a fixed budget and a prioritized backlog.

Implementing continuous planning and road mapping with agile portfolio management involves lighter-weight, more frequent planning cycles rather than heavy annual planning events, including:

- Quarterly portfolio roadmap reviews to adjust priorities based on the latest information.

- Bi-weekly or monthly portfolio Kanban reviews to address impediments and allocate resources.

- Ongoing backlog refinement and re-prioritization at the team level.

New metrics to track and measure the performance of an agile portfolio that emphasizes value delivery, speed, and quality over adherence to plans and budgets.

Metrics could include:

- Lead time and cycle time to measure the speed of value delivery.

- Value points delivered to quantify the business benefits realized.

- Net Promoter Score (NPS) to gauge customer satisfaction and advocacy.

- Agile maturity assessments to track progress on the agile transformation journey.

Implementing these practices requires a close partnership between the PMO, finance, business leaders, and agile teams. It's not a one-time event but an ongoing learning, adaptation, and continuous improvement journey.

The PMO's Role in Enabling Agile Portfolio Management

As the steward of the project portfolio, the PMO plays a crucial role in leading the transition to agile portfolio management, including:

Setting the Vision and Strategy: The PMO articulates a clear and compelling vision for what agile portfolio management will look like in the organization and how it aligns with broader business strategy by working closely with senior leaders to define strategic themes, objectives, and success measures.

Designing and Implementing Processes: The PMO will lead the design and rollout of new agile portfolio processes, such as lean budgeting, value stream organization, and continuous planning cycles through close collaboration with finance, IT, and business stakeholders to ensure processes are fit for purpose.

Upskilling and Coaching Teams: Agile portfolio management requires different skills and ways of working for project teams and business leaders. The PMO must invest in comprehensive training, coaching, and change management to build the necessary capabilities and mindsets.

Facilitating Governance and Decision-Making: While agile portfolio management emphasizes decentralized decision-making, there is still a need for lean governance and oversight at the portfolio level. The PMO facilitates portfolio review sessions, escalates issues and risks, and ensures strategic alignment.

Measuring and Communicating Performance: The PMO is responsible for defining, tracking, and reporting on agile portfolio metrics, including creating dashboards and information radiators, facilitating portfolio retrospectives, and communicating progress and outcomes to senior stakeholders.

Fostering a Culture of Agility: Perhaps most importantly, the PMO needs to lead the cultural transformation required for agile portfolio management to take hold. Modeling agile values, encouraging experimentation and learning, and creating a safe environment for teams to take risks and innovate are strongly suggested ways to engrain agile mindsets.

Agile portfolio management represents a significant shift from traditional ways of working, and it won't happen overnight. However, by incrementally introducing new practices, building the right capabilities, and fostering an agile culture, the PMO can guide the organization towards a more responsive, value-driven approach to managing its project investments.

Navigating Challenges and Pitfalls

As with any significant change initiative, the transition to agile portfolio management comes with challenges and potential pitfalls.

Some common issues that PMOs may encounter include:

- Resistance to change from stakeholders comfortable with traditional planning and funding models.

- Difficulty breaking down organizational silos and aligning around end-to-end value streams.

- Lack of clarity or alignment on strategic priorities and value measures.

- Challenges with implementing lean governance and decentralized decision-making.

- Inadequate training or coaching leads to inconsistent adoption of agile practices.

- Struggling to define and measure the right portfolio-level metrics.

- Reverting to old behaviors and mindsets under pressure.

Proactively anticipating and mitigating these challenges is a significant advantage for successful agile portfolio management implementation.

Strategies the PMO can employ include:

- Securing solid and authentic executive sponsorship and aligning the initiative with strategic goals.

- Starting with a pilot program or value stream to demonstrate proof-of-concept before scaling.

- Investing heavily in communication, education, and change management throughout the journey.

- Partnering closely with finance and business leaders to co-create new processes and metrics.

- Building a community of practice or center of excellence to support and sustain the transformation.

- Celebrating successes and sharing learnings early and often to build momentum and buy-in.

- Remaining open to feedback, iterating, and adapting the approach as needed based on experience.

Remember, agile portfolio management is an adventure of progress. The PMO can guide the organization towards a truly agile approach to managing its project portfolio by staying focused on the end goal of faster value delivery and customer responsiveness and persistently working through challenges and setbacks.

What Will You Do?

As you consider how to apply agile portfolio management principles and practices in your organization, here are some next steps to get started:

> 🎯 PMO Leaders: Educate yourself and your team on the fundamentals of lean portfolio management. Identify one or two practices you could pilot in the near term.
>
> 🎯 Program Managers: Work with business stakeholders to define the value streams in your domain. Consider how you could reorganize work and teams around these value streams.
>
> 📈 Project Managers: Evaluate your current project budgeting and funding processes and brainstorm ideas for making them leaner and more agile.
>
> 💼 Executives: Review your portfolio review and governance processes. Identify opportunities to streamline decision-making and empower teams.
>
> 👑 Chief Executives: Reflect on your organization's strategic priorities and how well your current portfolio aligns with them. Consider how agile portfolio management could help drive greater focus and responsiveness.

The shift to agile portfolio management is a significant undertaking. Still, the benefits - faster time-to-market, increased customer responsiveness, and more efficient resource allocation - can be game-changing. By aligning your portfolio with agile principles, implementing lean and agile practices, and leading the necessary organizational and cultural changes, you can position your PMO and your business for success in an increasingly dynamic and competitive landscape. The journey isn't easy, but for organizations committed to thriving in the age of agility, it's essential to embark upon it.

Chapter 5: Agile Governance and Leadership

As the PMO transforms to enable and support agility across the organization, one of the most critical shifts is how we approach governance and leadership. Traditional models of command-and-control, centralized decision-making, and heavy bureaucratic oversight are incompatible with the fast-paced, adaptive, and empowered nature of agile ways of working. Instead, the agile PMO must pioneer an innovative approach to governance and leadership that balances the need for strategic alignment and oversight with the autonomy and agility of empowered teams.

In this chapter, we'll explore agile governance and leadership in practice, the PMO's evolving role in enabling them, and strategies for overcoming common challenges and pitfalls.

Defining Agile Governance

At its core, governance is about ensuring that an organization directs its resources and makes decisions in alignment with its strategic goals while managing risk and ensuring compliance with relevant policies and regulations. In a traditional model, this often translates into heavy, top-down control mechanisms such as detailed project plans, strict change control processes, and multiple layers of management approvals.

Agile governance, in contrast, seeks to achieve strategic alignment and oversight through different principles and practices. Decentralized decision-making, with authority, is pushed down to the level of those closest to the information and the work. Rapid feedback loops and transparent information flow freely to enable real-time course correction and risk management. Lightweight, flexible policies and guardrails provide just enough structure without stifling innovation and adaptability. Collaborative, servant leadership empowers and enables teams rather than controlling them. Focus on outcomes and value delivery rather than adherence to plans and processes is at the forefront.

Agile governance creates the right conditions and guardrails for agile teams to operate effectively while providing the necessary strategic guidance and oversight to ensure they move in the right direction.

The PMO's Evolving Role in Agile Governance

As the project and program governance steward, the PMO is critical in defining and enabling agile governance across the organization. However, this requires a significant shift from the PMO's traditional functions and mindset.

Rather than acting as the "process police" enforcing compliance with rigid methodologies, the agile PMO becomes an enabler and coach, guiding teams towards agile best practices and helping them navigate obstacles and challenges. Key aspects of this new role include:

Defining Agile Policies and Guardrails: The PMO works with senior leaders and stakeholders to define the overarching policies, principles, and guidelines governing agile ways of working across the organization, such as:

- Agile methodologies and frameworks used (e.g., Scrum, Kanban, SAFe).

- Role definitions and responsibilities (e.g., Product Owner, Scrum Master).

- Metrics and KPIs for measuring agile performance and value delivery.

- Guardrails for budgeting, resource allocation, and risk management.

The key is to keep these policies lightweight and flexible, focusing on the minimum viable bureaucracy required to enable agile teams to operate effectively.

Coaching and Enabling Agile Teams: Rather than directly controlling projects, the PMO shifts to a coaching and enabling role, providing guidance, training, and support to help agile teams continuously improve their practices and deliver value.

In its enablement role, the PMO ensures the following:

- Training and certifying team members in agile methodologies and tools.

- Facilitating agile ceremonies and retrospectives.

- Helping teams remove impediments and navigate organizational roadblocks.

- Connecting teams with subject matter experts and resources as needed.

The goal is to build self-sufficiency and mastery within teams rather than having them rely on the PMO for direction.

Facilitating Lean Governance Practices: While agile teams are empowered to make decentralized decisions, there is still a need for lean governance and oversight at the portfolio level. The PMO plays a pivotal role in facilitating practices such as:

- Regular portfolio review sessions to monitor progress, manage risks, and adjust priorities.

- Lean budgeting and funding processes that allocate resources dynamically based on value.

- Transparent reporting and information radiators to provide real-time visibility into agile performance.

- Escalation and issue resolution processes to address impediments and risks.

The PMO doesn't make decisions unilaterally but facilitates collaborative decision-making among stakeholders.

Cultivating Agile Leadership and Culture: Perhaps most crucially, the PMO influences leadership mindsets and cultural norms required for agility to thrive, such as:

- Coaching leaders at all levels to adopt a more servant leadership and empowerment mindset.

- Encouraging experimentation, learning from failure, and continuous improvement as cultural norms.

- Facilitating cross-functional collaboration and breaking down organizational silos.

28

⊙ Recognizing and celebrating agile successes and role models.

Cultural change is the most challenging aspect of an agile transformation and is critical for long-term sustainability.

Strategies for Effective Agile Governance

Implementing agile governance is not a one-time event but an ongoing learning, adaptation, and continuous improvement journey. Some key strategies the PMO can employ to make this transition effective include:

Start with the Why: Ensure a clear and compelling rationale for the shift to agile governance tied to strategic business goals. Communicate this "why" early and often to build understanding and buy-in.

Secure Executive Sponsorship: Agile governance requires top-down support and role modeling from senior leaders. The PMO needs to partner with executives to align on the vision and secure their active sponsorship.

Take an Incremental Approach: Rather than launching a wholesale governance overhaul, start with a few practices in a pilot area, learn and iterate, and then gradually scale what works. This incremental approach helps build momentum and mitigate risk.

Invest in Education and Enablement: Extensive training, coaching, and hands-on support are needed to help leaders and teams adopt new agile governance mindsets and practices. The PMO should invest heavily in building these capabilities.

Measure and Communicate Value: Define clear metrics and KPIs upfront to track the impact of agile governance and regularly communicate progress and successes to stakeholders to build and sustain momentum.

Embrace a Learning Mindset: Recognize that the transition to agile governance will involve challenges, missteps, and learnings. Foster a culture of experimentation, feedback, and continuous improvement.

By employing these strategies, the PMO can effectively guide the organization through the transition to agile governance while modeling the agile principles it seeks to instill.

The Importance of Servant Leadership

One of the most fundamental mindset shifts required for agile governance is transitioning from a command-and-control leadership model to a servant leadership approach.

In a servant leadership model, the traditional hierarchy turns upside down. Instead of teams serving leaders, leaders exist to serve teams. Their primary role is to remove obstacles, provide support and resources, and enable teams to do their best work.

Characteristics of servant leaders include:

- Empathy and listening to understand team members' perspectives and needs truly.

- Stewardship and a commitment to serving the greater good of the organization.

- The ability to anticipate risks and opportunities.

- Persuasion and influence rather than authority to build alignment and commitment.

- Conceptualization and the ability to dream big while also being grounded in day-to-day realities.

- Commitment to the growth and development of people and teams

- Building connections that create a sense of belonging.

For the PMO, embracing and modeling servant leadership is essential to enabling empowered, self-managing teams, which are the hallmarks of agility by letting go of the need for control, trusting teams to make good decisions, and focusing on enabling rather than directing.

It's a significant shift that requires personal reflection, empathy, humility, and openness to learning and growth. However, for PMO leaders who embrace this approach, the rewards can be truly transformational regarding team engagement, innovation, and business impact.

What Will You Do?

As you contemplate the shift to agile governance and leadership in your organization, consider the following steps:

> 🎯 PMO Leaders: Assess your current governance policies and processes through an agile lens. Identify areas where you can streamline, simplify, or empower teams.
>
> 🚀 Program Managers: Reflect on your leadership style and identify opportunities for a more servant-leadership approach. Commit to one specific action to enable and empower your team.
>
> 📈 Project Managers: Evaluate the metrics and KPIs you use to measure project success. Consider how you could shift to more agile measures focused on value delivery and customer satisfaction.
>
> 💼 Executives: Reflect on the tone you set for governance and oversight in your organization. Identify ways you could visibly role model agile principles and empower teams.
>
> 👑 Chief Executives: Consider your organization's cultural norms and leadership behaviors today. What shifts might be needed to foster greater agility, and how will you lead that change?

The transition to agile governance and leadership is a journey that requires commitment, perseverance, and a willingness to learn and adapt continuously. It's not always easy, but the benefits—speed, innovation, engagement, and value delivery—can be game-changing for organizations that get it right. By stepping up to lead this transformation and role-modeling agile principles every step of the way, the PMO can be a true catalyst for organizational success in the age of agility.

Chapter 6: Agile Performance Measurement and Reporting

As the PMO shepherds the organization through an agile transformation, one of the most critical areas of change is how we measure and report on performance. Traditional project metrics, heavily focused on schedule, budget, and deliverable completion, are insufficient in an agile world. They need to capture what truly matters: customer satisfaction with delivery.

In this chapter, we'll explore why a new approach to performance measurement matters to an agile organization and how the PMO can implement agile reporting practices to drive better decision-making and continuous improvement.

The Need for a New Approach

In a traditional project management environment, we measure success by the triple constraint: delivering a defined scope on schedule and within budget. While these dimensions are important, they reflect several flawed assumptions:

- That scope is defined and locked in upfront.

- Delivering the original plan equates to providing value.

- That value is only created and measured at the end of a project.

Agile approaches challenge these assumptions. They recognize that the scope must be flexible and adaptable in a complex and rapidly changing world based on continuous feedback and learning. They assert that the ultimate measure of success is not adherence to a plan but delivering value to customers. They shift the focus from big-bang, end-of-project value to incremental value delivered continuously throughout the project lifecycle.

As such, agile demands a new approach to performance measurement that aligns with these principles and enables real-time insights into value delivery.

Metrics That Matter in an Agile World

What metrics matter in an agile organization? While specific KPIs will vary based on the context and goals of each organization, some key areas to focus on include:

Value Delivery: The ultimate measure of agile success is the continuous delivery of value to customers. Metrics in this area might include:

- Business value points delivered per iteration or release.

- Net Promoter Score (NPS) or customer satisfaction ratings.

- Revenue generated or costs saved from delivered features.

- Number of live customers using delivered features.

Speed and Predictability: Agile aims to deliver value faster and more predictably.

Key metrics here include:

- Lead time (the time from idea to delivery)

- Cycle time (the time from starting work on a feature to delivering it).

- Throughput (the number of features or story points delivered per unit of time).

- On-time delivery rate (the percentage of committed features delivered per iteration or release).

Quality and Sustainability: Agile emphasizes building in quality and creating a sustainable pace of work. Relevant metrics include:

- Defect escape rate (the number of defects found in production).

- Technical debt (the cost of reworking low-quality code).

- Automated test coverage (the percentage of code covered by automated tests).

- Team happiness or engagement scores.

Continuous Improvement: A fundamental tenet of agile is the continuous improvement of processes and practices.

Metrics to track this include:

- Velocity trend over time (the trend of story points delivered per iteration).

- Cycle time trend over time (the trend of the time to deliver a feature).

- Number of process improvements or experiments run per quarter.

- Team retro scores (the team's self-assessment of their performance and improvement).

These are just a sampling of potential agile metrics—the specific KPIs that matter most will depend on your organization's unique context and goals. Metrics should relate to customer value delivery and provide actionable insights for continuous improvement.

Implementing Agile Reporting Practices

Measuring the right things is just the first step - equally important is implementing reporting practices that enable real-time, actionable insights. Some practices the PMO can implement include:

Automate Data Collection: Manually tracking and compiling agile metrics is time-consuming and error-prone. Invest in tools that automate data collection directly from agile work management systems (e.g., Jira, Azure DevOps) to ensure accurate and real-time data.

Create Visual Dashboards: Agile metrics should be visible and accessible to all stakeholders. Create visual dashboards that display key metrics in an easy-to-understand format and make them available on large screens, web portals, or mobile apps.

Report Trends, Not Just Snapshots: Agile is about continuous improvement over time. Ensure your reporting focuses on trends and changes in key metrics, not just single-point-in-time snapshots.

Integrate with Planning and Retro Processes: Agile reporting should not be a separate, after-the-fact activity. Integrate reporting into iteration planning, daily stand-ups, and retrospectives to enable real-time decision-making and process improvement.

Focus on Action, Not Just Insight: Reporting is only valuable if it drives action. Ensure each metric ties to clear owners and action plans. Use reporting as a launch pad for problem-solving and continuous improvement efforts.

Tailor to the Audience: Different stakeholders need various levels of detail and focus. Create tailored views or reports for executive sponsors, product owners, development teams, and other key audiences.

By implementing these practices, the PMO can create a reporting ecosystem that provides real-time, actionable insights into agile performance and drives continuous organizational improvement.

The PMO's Role in Agile Performance Measurement

As the steward of project performance, the PMO plays a critical role in defining, implementing, and continuously improving agile measurement practices across the organization. Key aspects of this role include:

Defining the Agile Metrics Strategy: The PMO works with executive sponsors, product owners, and agile teams to define the key metrics that they will use to measure agile success. To make this a reality, the PMO could hold workshops to identify and align on value drivers, select specific KPIs, and determine targets and benchmarks.

Implementing Agile Reporting Tools and Processes: The PMO leads the selection, configuration, and rollout of agile reporting tools and dashboards, including source-of-truth data, reporting cadence, and formats.

Training and Enabling Teams: Agile reporting is a significant shift for many organizations. The PMO provides training, coaching, and support to help teams understand and utilize the new metrics and reporting practices in their day-to-day work.

Facilitating Action and Improvement: Most critically, the PMO facilitates continuous improvement based on agile reporting insights.

This includes leading problem-solving workshops, conducting root-cause analyses, and driving action plans to address performance gaps.

Communicating Value to Stakeholders: The PMO is the conduit for communicating agile performance to executive sponsors and stakeholders. As such, the PMO creates executive-level dashboards, presenting key insights and trends and telling the story of how agile delivers business value.

By taking on this strategic and enabling role, the PMO becomes the central hub for agile performance insights across the organization, driving better decision-making, continuous improvement, and business value.

Overcoming Challenges and Pitfalls

As with any significant change, implementing agile performance measurement has its share of challenges and potential pitfalls. Some common issues the PMO may face include:

- Resistance to new metrics from teams comfortable with traditional measures.

- Over-focusing on easy-to-measure but less meaningful metrics (e.g., velocity).

- Lack of data quality or consistency across teams and tools.

- Reporting for reporting's sake, without driving action or improvement.

- Misusing metrics to micromanage or punish teams rather than enable them.

Proactively anticipating and addressing these challenges helps facilitate adoption. Some strategies the PMO can employ include:

- Securing strong and authentic executive sponsorship and aligning metrics to strategic goals.

- Engaging teams early and often in defining and refining metrics.

- Investing in data quality and governance to ensure trust in the numbers.

- Focusing on a small set of actionable metrics rather than measuring everything.

- Emphasizing trends and improvement over hitting arbitrary targets.

- Using metrics to enable and empower teams, not to control or punish.

Remember, agile performance measurement aims not to create a perfect set of KPIs but to enable a continuous learning and improvement culture focused on delivering customer value.

What Will You Do?

As you contemplate implementing agile performance measurement in your organization, consider the following steps:

> 🎯 PMO Leaders: Facilitate a workshop with key stakeholders to align on the value drivers and metrics that matter most for your agile initiatives.
>
> 🐀 Program Managers: Work with your agile teams to identify how to use metrics to improve their performance sprint over sprint.
>
> 📈 Project Managers: Assess your current project reporting practices. Identify reports or metrics that no longer add value in an agile world.
>
> 💼 Executives: Reflect on the metrics you currently use to assess project health and success. Consider how you might shift your focus to align with agile principles.
>
> 👑 Chief Executives: Evaluate the performance measurement culture in your organization. Is it focused on enabling and empowering teams or on controlling and micromanaging? What shifts might be needed?

The transition to agile performance measurement is a journey, not a destination. By focusing on the metrics that truly matter, implementing actionable reporting practices, and using insights to drive continuous improvement, the PMO can help guide the organization toward a more value-driven, customer-centric, and successful approach to performance management. The unique position of PMO leaders enables them to drive the overall agile transformation.

Chapter 7: Scaling Agile across the Organization

As organizations reap the benefits of agile at the team level - faster delivery, higher quality, improved customer satisfaction - the logical next question becomes, "How do we scale these benefits across the entire enterprise?" Indeed, for organizations to truly achieve business agility, agile principles and practices need to extend beyond isolated pockets of innovation and become the default way of working across functions, geographies, and levels of the hierarchy.

In this chapter, we will explore what it means to scale agile, key challenges and considerations, and the PMO's critical role in enabling and supporting the scaling journey.

Understanding Agile at Scale

Scaling agile is about more than having many agile teams across the organization. It is about aligning and coordinating the work of those teams toward strategic business objectives and creating an organizational system that enables fast decision-making, rapid learning, and continuous improvement at all levels.

Fundamental aspects of agile at scale include:

Aligning Teams to Value Streams: In a scaled agile organization, teams organize around value streams - end-to-end processes that deliver value to customers. This shift requires breaking down silos and creating cross-functional, highly collaborative teams.

Implementing Lean Portfolio Management: Scaled agile organizations need an approach to managing the portfolio of work. Lean Portfolio Management (LPM) applies lean and systems thinking to strategy, investment funding, governance, and agile program guidance.

Enabling Continuous Delivery: Agile at scale requires rapidly and continuously delivering value to customers. The new process involves establishing a continuous delivery pipeline with automated testing, integration, and deployment capabilities.

Fostering Continuous Learning and Improvement: Scaling agile is not a one-time event. Creating an environment where learning, flexibility, and adaptability become the norm will significantly improve the chance of success.

Developing Agile Leadership and Culture: Most importantly, scaling agile requires a fundamental leadership mindset and organizational culture shift. Leaders must let go of command-and-control behaviors and embrace a more empowering servant leadership style.

Scaling frameworks like Scaled Agile Framework (SAFe), Large-Scale Scrum (LeSS), and Disciplined Agile (DA) provide structured approaches for implementing these elements. However, it is important to remember that scaling agile is not about rigidly following a framework but adapting principles to fit your unique organizational context.

The PMO's Role in Enabling Agile at Scale

The PMO drives and supports the agile scaling journey across the organization. As the central hub for strategic planning, governance, and delivery enablement, the PMO can help create the conditions for agility to thrive at scale.

The PMO's role in scaling agile includes:

Defining the Agile Scaling Strategy: The PMO works with executive leadership to determine the "why, what, and how" of agile scaling. This critical initial step includes aligning executive leader expectations, articulating the business case, selecting the scaling approach, and defining the roadmap and milestones.

Designing the Agile Operating Model: The PMO leads the design of the new agile operating model, including organizational structure, governance mechanisms, funding processes, and delivery practices, which includes close collaboration with Finance, HR, IT, and other business functions.

Enabling Lean Portfolio Management: The PMO implements LPM practices, including defining value streams, establishing lean budgets and guardrails, and facilitating portfolio-level planning and governance.

Building Agile Capabilities: Scaling agile requires new organizational skills and competencies. The PMO leads the effort to assess capability gaps, define learning paths, and deliver training and coaching to build agile mastery.

Fostering Continuous Improvement: The PMO establishes the cadence and practices for continuous improvement at scale, including agile maturity assessments, portfolio retrospectives, and improvement workshops.

Measuring and Communicating Progress: The PMO defines and tracks key metrics to measure the progress and impact of the agile scaling effort and regularly communicates status and learnings to key stakeholders.

The PMO becomes the central catalyst for driving agile transformation at scale by taking on this strategic enabling role.

Overcoming Challenges and Roadblocks

Scaling agile has its challenges. Common roadblocks organizations face include the following:

- Resistance to change from middle management who may feel threatened by the shift to self-organizing teams.
- Difficulty breaking down silos and aligning around end-to-end value streams.
- Lack of executive sponsorship and vision for the agile scaling effort.
- Misalignment between agile delivery and traditional HR, Finance, and governance processes.
- Inadequate investment in training and coaching to build new agile capabilities.
- Falling back into old command-and-control habits under pressure.

The PMO can help mitigate these challenges by:

- Securing executive sponsorship and aligning the scaling effort to strategic business goals.

- Engaging middle managers early and often, helping them define their new roles in an agile model.

- Collaborating closely with HR, Finance, and other functions to redesign misaligned processes.

- Investing heavily in change management, communication, and capability-building efforts.

- Establishing governance mechanisms to surface and resolve impediments quickly.

- Leading by example and role modeling agile behaviors even in the face of pressure.

Scaling agile is a significant undertaking that requires persistence. But for organizations that get it right, the benefits - faster innovation, higher employee engagement, and improved business results - can be genuinely transformative.

Scaling Agile Governance

As organizations scale agile, a key challenge is maintaining the benefits of agile self-organization while ensuring alignment and governance across the enterprise. Left unchecked, a proliferation of autonomous agile teams can lead to duplication of effort, misalignment with strategic priorities, and a lack of enterprise-wide standards and discipline.

The PMO is critical in striking the right balance between empowered execution and strategic alignment.

Practices for scaled agile governance include:

Defining Lean Guardrails: Rather than imposing heavy top-down controls, the PMO establishes a set of lean guardrails - minimum standards, policies, and constraints within which agile teams must operate. These include architectural standards, security and compliance requirements, and investment horizons.

Implementing Agile Portfolio Management: The PMO facilitates regular portfolio planning and review sessions to set strategic priorities, identify how to fund value streams, and collaborate on managing dependencies. This provides the high-level context and alignment within which agile teams execute.

Enabling Decentralized Decision-Making: The PMO empowers agile teams to make decentralized decisions within the defined guardrails and strategic priorities. Decisions around delivering value are pushed down to those closest to the information, with fast feedback loops to course correct as needed.

Establishing Lean Governance Forums: The PMO sets up lightweight governance forums at the program and portfolio level, focused on removing impediments, managing risk, and ensuring alignment. These forums rotate a cast of agile team members to ensure informed decision-making with on-the-ground context.

Measuring and Monitoring Outcomes: The PMO focuses on measuring and monitoring business outcomes rather than tightly controlling inputs and outputs. Key Value Stream and program-level KPIs are defined, with regular review cycles to assess progress and make data-driven decisions.

Fostering Continuous Improvement: The PMO facilitates regular retrospectives and improvement workshops at the program and portfolio level to identify systemic impediments and drive continuous improvement across value streams.

By implementing these practices, the PMO can create a governance model that provides enough alignment and control while enabling the benefits of agile self-organization.

The Journey to Business Agility

The goal of scaling agile is to have more agile teams and achieve true business agility—the ability to sense and effectively respond to changes rapidly. Holistic transformation beyond delivery practices encompasses strategy, structure, processes, people, and technology.

The PMO orchestrates this enterprise-wide transformation in close partnership with executive leadership and functional stakeholders. By defining the North Star vision, designing the operating model, building capabilities, and fostering a culture of continuous improvement, the PMO can help steer the organization toward true business agility.

It is a journey that requires significant commitment, flexibility, and patience. For organizations that persist, the rewards are game-changing.

What Will You Do?

As you consider the agile scaling journey in your organization, here are some next steps to get started:

> 🎯 PMO Leaders: Educate yourself and your team on agile scaling frameworks and practices. Consider which approach might best fit your organizational context and culture.
>
> 💼 Program Managers: Assess your current program structure and practices through an agile lens. Identify opportunities to organize around value streams and enable authentic self-organization.
>
> 📈 Project Managers: Develop your skills and knowledge in scaled agile practices, such as Agile Release Train facilitation, Lean Portfolio Management, and DevOps.
>
> 💼 Executives: Engage your PMO leader in a discussion about your organization's readiness and appetite for agile scaling. Identify the potential business benefits and executive sponsorship needed.
>
> 👑 Chief Executives: Reflect on your organization's current level of business agility. Consider how scaling agile might help you better respond to market changes and customer needs.

Remember, scaling agile is not a destination but a continuous learning and improvement journey. By starting small, learning fast, and adapting as you go, you can begin to unleash the power of agility across your enterprise. The PMO's role in this journey is crucial - not just as a facilitator of practices but as a strategic enabler of business transformation. It is a challenging but gratifying undertaking that will position your organization for success in the fast-paced, ever-changing digital world.

Chapter 8: The Future of Agile PMOs

As we have explored throughout this book, the rise of agile transforms how we deliver projects and think about the nature of work and value creation in organizations. For PMOs, this shift represents both a challenge and an opportunity. It requires letting go of long-held beliefs and practices around command-and-control, predictability, and linear planning and embracing a new role as enablers of organizational agility, innovation, and continuous learning.

In this closing chapter, we'll examine the future of Agile PMOs. We will explore emerging trends and practices, the impact of digital disruption, and the skills and capabilities that will define PMO success in the future. We will also provide some practical guidance on how PMOs can continue to evolve and mature their agile practices over time.

Emerging Trends and Practices

The agile world constantly evolves, with new frameworks, techniques, and technologies always emerging. While it is impossible to predict the future with certainty, here are a few key trends and practices that are shaping the direction of Agile PMOs:

Beyond Software Development: Agile began in software development, but its principles and practices are now applied far beyond IT. Organizations recognize the benefits of agility in all functions, from marketing to HR to finance. PMOs must adapt their approaches to support agile in these diverse contexts.

Scaling Agile to the Enterprise: As discussed in the previous chapter, scaling agile across the enterprise is a key challenge and opportunity for PMOs. Emerging frameworks like SAFe, LeSS, and DA provide structured approaches, but the real work is to tailor these to your organization's unique context and culture.

The Rise of DevOps and Continuous Delivery: Agile is increasingly intertwined with DevOps, integrating development and operations to deliver continuous value. To support this shift, PMOs must build automation, infrastructure-as-code, and site reliability engineering capabilities.

Lean and Agile Portfolio Management: Traditional and continuous portfolio management approaches. Lean Portfolio Management (LPM) applies agile and lean principles to strategy and investment funding for greater agility and responsiveness.

Design Thinking and Customer Centricity: Agile is fundamentally about delivering customer value. Practices like design thinking, customer journey mapping, and continuous user feedback are becoming integral to agile delivery. PMOs will need to build capabilities in these areas to drive customer-centric innovation.

The Agile Organization: Agile is moving beyond a set of delivery practices to a holistic organizational operating model and involves rethinking structures, governance mechanisms, HR policies, and leadership behaviors to enable enterprise-wide agility. PMOs will play a key role in designing and facilitating this transformation.

These are just a few examples of the many trends and practices shaping the future of Agile PMOs. The key is to stay attuned to these developments, experiment with novel approaches, and continuously adapt and improve over time.

The Impact of Digital Disruption

The relentless pace of digital disruption is the most significant factor shaping the future of Agile PMOs. Advances in areas like artificial intelligence, blockchain, the Internet of Things, and quantum computing are upending traditional business models and reshaping entire industries. In this context, the ability to innovate, experiment, and pivot rapidly is not just a nice-to-have—it's a matter of survival.

For PMOs, digital disruption presents both challenges and opportunities:

- The need for speed and agility is more significant than ever before. PMOs will need to help organizations deliver value faster, more frequently, and reliably to keep pace with the rate of change.

- Customer expectations can outpace reality and practicality. PMOs must help organizations deeply understand and empathize with customers and rapidly prototype and iterate on innovative solutions to meet their ever-evolving needs.

New technologies are enabling new ways of working. PMOs must stay on top of emerging tools and platforms and help the organization harness them for collaboration, automation, and continuous improvement.

To thrive in digital disruption, PMOs must implement agile practices and embody agile values and principles in all aspects of their work by scanning the horizon for new opportunities and threats and proactively steering the organization to respond. It also means fostering a culture of experimentation, learning, and resilience in constant change.

The Skills and Capabilities of Future PMOs

To enable and support the agile organization of the future, PMOs will need to develop a new set of skills and capabilities, including:

Business Acumen: PMOs deeply understand the business context, strategic objectives, and competitive landscape. They will need to speak the language of business value and outcomes, not just project deliverables and milestones.

Agile and Lean Mastery: PMOs have expertise in agile and lean principles, practices, and frameworks, adapting and tailoring them to different contexts and continuously evolving and improving them over time.

Data and Analytics Savvy: PMOs are skilled in data analysis, visualization, and storytelling. They will need to be able to leverage data to drive insights, decisions, and continuous improvement.

Design and Innovation Skills: PMOs must be adept at human-centered design, ideation, and prototyping to foster organizational creativity and innovation and help teams rapidly experiment with new solutions.

Change Leadership: PMOs will need to be skilled change agents, able to drive and support transformation at the individual, team, and organizational levels. They will need to be adept at influencing, coaching, and inspiring others to embrace new ways of working.

Emotional Intelligence: A self-aware PMO displays empathy and social skills to build trust, resolve conflicts, and foster collaboration across diverse teams and stakeholders.

Continuous Learning: PMOs are constant learners, constantly seeking new knowledge, skills, and perspectives. They'll need to model curiosity, humility, and a growth mindset and create a learning culture across the organization.

Developing these capabilities will require significant training, coaching, and experiential learning investment for PMO staff. It will also require rethinking traditional PMO roles and structures and creating new career paths and development opportunities aligned with the skills of the future.

Measuring and Maturing Agile PMO Capabilities

As Agile PMOs evolve and mature their practices over time, it's essential to have a way to measure and assess progress. Agile maturity models provide a structured approach to evaluating capabilities across organizational dimensions, identifying areas of strength and opportunity, and prioritizing improvement efforts.

Standard dimensions of Agile PMO maturity include:

- Agile Practices: The extent to which agile practices like iterative delivery, cross-functional teams, and adoption of continuous improvement across the organization.

- Lean Portfolio Management: The maturity of practices for aligning strategy, funding, and execution around value streams and enabling fast, flexible value delivery.

- DevOps and Continuous Delivery: The extent to which automated and integrated development and operations enable frequent, reliable releases to production.

- Data and Analytics: The maturity of practices for leveraging data to drive insights, decisions, and actions across the portfolio.

- Organizational Agility: The extent to which agile values and mindsets permeate the organization's culture, structure, and leadership.

For each dimension, the maturity model defines a set of progressive levels (e.g., from "Initial" to "Optimizing"), with clear criteria and indicators for each level. PMOs can continuously evolve and advance their agile capabilities by regularly assessing maturity levels, identifying gaps, and defining improvement roadmaps.

It is important to remember that maturity is not the end goal—the aim is better business outcomes and customer value. Maturity models are a guide and a conversation starter, not a rigid prescription or a box-ticking exercise. The real work is continuously experimenting, learning, and adapting to find what works best for your unique organization and context.

The Agile PMO as a Learning Organization

Ultimately, the future of Agile PMOs is about becoming genuine learning organizations that constantly sensitize, respond, and adapt to change. In a world of rapid disruption and uncertainty, the ability to pivot more efficiently than the competition is the ultimate sustainable advantage.

Creating a continuous learning and improvement culture for PMOs at all levels means fostering curiosity, experimentation, and knowledge-sharing across teams and functions. It means leveraging data and insights to refine and optimize practices. And it means embracing failure as a learning opportunity and not an adverse event to avoid at all costs.

Becoming a learning organization is not a one-time event but an ongoing journey. It requires leaders who model humility, vulnerability, and a growth mindset. It requires structures and processes that enable rapid feedback, reflection, and adaptation. It requires a deep commitment to investing in the growth and development of people—not just their technical skills but their creativity, critical thinking, and emotional intelligence.

For PMOs that embrace this learning mindset, the possibilities are endless. They can become enablers of agile delivery and drivers of organizational transformation and innovation. They can help their organizations not just survive but thrive in the face of constant change and disruption. And they can create a lasting legacy of value creation and positive impact for all stakeholders.

What Will You Do?

As we conclude this exploration of the Agile PMO, we invite you to reflect on your learning journey:

> 🎯 PMO Leaders: Assess your current PMO capabilities against an agile maturity model. Identify one or two areas for improvement, and develop a plan to advance your practices.
>
> 🏹 Program Managers: Reflect on how well your current program embodies agile values and principles. Identify one behavior or mindset you want to work on personally.
>
> 📈 Project Managers: Consider how you can leverage data and analytics to drive more insights and continuous improvement in your projects. Commit to experimenting with one new data-driven practice.
>
> 💼 Executives: Evaluate how well your current organizational structure and governance enable agility. Identify one key barrier to remove or enabler to implement.
>
> 👑 Chief Executives: Reflect on your leadership style and how it models learning, humility, and agility. Commit to one behavior change to inspire others.

Remember, the journey to agility is never complete. It is a constant process of learning, experimentation, and growth. By embracing this mindset and continuously evolving your practices, you can help your organization thrive in a world of constant change.

The Agile PMO has a crucial role in this journey—as a project facilitator and a strategic partner in driving organizational transformation. It is a challenging but gratifying role that combines technical skills, business acumen, and leadership capabilities.

Stay curious, humble, and committed to continuous learning and growth as you continue your Agile PMO journey. The future is uncertain, but the possibilities are endless, with agility as your guide.

www.ingramcontent.com/pod-product-compliance
Lightning Source LLC
Chambersburg PA
CBHW041719200326
41520CB00005B/221